AF897727

The Food System

REVOLUTION

Shannon J Dobbs

TABLE OF CONTENTS

Introduction

THE FOOD SYSTEM CRISIS AND THE POWER OF MARKET-ALIGNED SOLUTIONS

Imagine a world where grocery stores overflow with fresh produce, while just blocks away, families struggle to put nutritious meals on the table. Picture farmers forced to let crops rot in fields, while dumpsters behind restaurants brim with uneaten food. This isn't a dystopian fantasy—it's the reality of our modern food system.

We live in a world of paradoxes. We produce more than enough food to feed everyone on the planet, yet millions go hungry. We've never been more aware of the importance of nutrition, yet diet-related diseases are skyrocketing. We're more concerned than ever about climate change, yet our food system continues to be one of the largest contributors to greenhouse gas emissions.

These aren't just abstract problems—they're crises that affect real people every day. I've seen it firsthand. As a nightclub owner in Reno, Nevada, I watched my customers—my friends—struggle daily with accessing healthy food. And when I found myself battling preventable, food-related diseases, I realized how lucky I was to have the resources to make changes. Many aren't so fortunate.

This personal wake-up call led me on a journey that's transformed my understanding of our food system—and my vision for how we can fix it. What I've discovered is both alarming and inspiring. Alarming because the scale of the problem is even bigger than most realize. Inspiring because the solutions are within our grasp—if we're willing to think differently.

Here's a startling fact: **over a third of all climate change-causing greenhouse gas emissions are tied to our food system.** From agriculture to transportation, from processing to waste, our current approach to food is **cooking our planet.** But here's the flip side: by transforming our food system, we have an enormous opportunity to combat climate change, improve public health, and create more equitable communities.

The challenge is that our current approaches aren't cutting it. Traditional charity models, while wellintentioned, often treat symptoms rather than causes. They're not designed to **create lasting change** or operate at the scale needed to address these systemic issues. And while policy changes are important, they often move at a glacial pace compared to the urgency of our challenges.

That's where market-aligned strategies come in. By harnessing the efficiency and scalability of market mechanisms—but directing them towards social and environmental goals—we can create solutions that are both powerful and sustainable. We can build systems that **nourish communities, support farmers, rescue food from waste, and fight climate change**—all while generating the resources needed to **sustain and expand** these efforts.

This isn't just theory. Throughout this book, you'll learn about innovative models already making a difference. You'll discover how reimagined grocery stores are bringing healthy food to underserved communities. You'll explore how "food rescue hubs" are transforming would-be waste into valuable resources. You'll see how blockchain technology can create more transparent and efficient food systems.

But this book isn't just about big ideas—it's a call to action. Whether you're a community leader, an entrepreneur, a policymaker, or simply someone who cares about these issues, you have a role to play in this food system revolution. You'll find practical strategies, real-world case studies, and concrete steps you can take to make a difference in your community.

We stand at a critical juncture. The challenges we face are enormous, but so is our potential to create change. By **reimagining our relationship with food**—how we grow it, distribute it, consume it, and think about it—we can build a more **just, sustainable, and nourishing world.**

This revolution isn't about destroying our current system—it's about transforming it from within, using the power of markets **to serve people and the planet**. It's about creating a food system that **works for everyone**, not just a privileged few. It's about seeing the connections between our plates, our communities, and our planet—and taking action to strengthen those connections.

As we embark on this journey together, I invite you to open your mind to new possibilities. Challenge your assumptions about how things "should" work. And most importantly, recognize your own power to create change. The future of our food system—and our planet—depends on it.

Are you ready to join the food system revolution? Let's dig in.

The greatest opportunity of our time lies not in creating something new, but in reimagining what already exists.

Chapter 1

UNDERSTANDING THE FOOD SYSTEM CRISIS

"

Standing behind the bar at 5 Star Saloon in downtown Reno, I watched as my friends and customers - people I cared about deeply - struggled with their health. One friend battled congestive heart failure, another lost part of his hand to diabetes complications before succumbing to the disease. These weren't just statistics; they were people I knew, laughed with, and worked alongside. These tragedies were a wake-up call that our food system is failing us, and in the most personal way possible.

To understand how we arrived at this point—and more importantly, how we can chart a course to a better future—we need to take a hard look at our current food system. It's a complex web of producers, distributors, retailers, and consumers, each playing a role in getting food from farm to fork. But this system, designed for efficiency and profit, has critical gaps where food and people fail to connect.

The State of Our Plates

Let's start with some sobering statistics. In the United States, a country of plenty:

- Over 38 million people, including 12 million children, are food insecure.
- Meanwhile, 40% of all food produced goes to waste.
- Diet-related diseases like obesity and diabetes have reached epidemic proportions, affecting both our national health and our economy.

These numbers aren't just data points—they represent real people struggling with hunger, farmers watching their livelihoods wither, and communities grappling with the health consequences of a broken food system.

The rise of food deserts—areas where residents lack access to affordable, nutritious food—is a stark illustration of this system's failings. Imagine living in a place where fresh produce is as rare as an oasis in the desert. For millions of Americans, this is reality. These food deserts are disproportionately found in low-income neighborhoods and communities of color, perpetuating cycles of poor health and economic disadvantage.

But the crisis extends beyond individual health. In cities like Reno civic leaders performatively scratch their heads, publicly wondering why they can't attract a grocery store downtown by not so subtly name dropping their favorite high profile chains while rejecting grassroots efforts wholesale. But these stores operate on complex algorithms to determine where to place stores and there's a financial reason these food deserts exist. The answer lies not in chasing big-name investment, but in understanding the complex web of our food system and the communities it serves - or fails to serve.

The Hidden Climate Impact

But the crisis doesn't stop at our dinner tables. Our food system is quietly fueling one of the greatest challenges of our time: climate change. Here's a startling fact that's often overlooked: 35% of climate change-causing greenhouse gas emissions, over a third of the annual total, can be traced back to our food system.

This impact is often hidden, fragmented across various sectors in climate reports. Take a look at this chart of Colorado's greenhouse gas emissions [insert image of emissions chart]. At first glance, agriculture seems to play a minor role. But this visualization is deeply misleading. The reality is that our food system's impact is buried across multiple categories—from transportation and industrial energy use to waste management.

When we connect these dots, we uncover a startling truth: our food system is a major driver of climate change. From the methane produced by food decomposing in landfills to the carbon emissions from transporting food across long distances, the environmental cost of our current approach to food is staggering.

This fragmented view of our food system's impact isn't unique to Colorado. It's a nationwide blind spot that obscures the true scale of the problem - and the potential for transformative solutions. By connecting these dots, we uncover not just a challenge, but an opportunity for systemic change.

The Paradox of Abundance and Scarcity

Perhaps the most perplexing aspect of our current food system is the coexistence of abundance and scarcity. The United States produces more than enough food to feed everyone. Yet millions go hungry, while billions of pounds of food rot in landfills.

This paradox points to a fundamental problem: it's not a lack of food that's the issue, but how that food moves—or doesn't move—from farm to fork. Our current system is riddled with inefficiencies and misaligned incentives that result in massive waste and unequal access.

For example:

- Aesthetic standards in supermarkets lead to the rejection of perfectly edible but "ugly" produce.
- Rigid company mandated sell-by dates result in the disposal of food that's still safe to eat.
- Lack of infrastructure for food recovery means that excess prepared food often ends up in the trash rather than on the plates of those who need it.

These are not insurmountable problems. They're the result of a system designed without consideration for its full impact—on people, on communities, and on the planet.

I saw this paradox play out daily all over Nevada for nearly 25 years. Just blocks from casinos overflowing with lavish buffets, families struggled to put nutritious meals on the table. It's not just a failure of distribution; it's a failure of imagination and will.

The Misconception of Simple Solutions

City officials and developers often think attracting a big-name grocery store will solve food access issues. But parachuting in a supermarket isn't a silver bullet. It doesn't address the root causes of food insecurity or build community resilience. We need to think bigger - and sometimes, paradoxically, smaller and more local.

The Urgency of Now

The consequences of our broken food system extend far beyond empty stomachs and overflowing landfills. They touch every aspect of our society:

- Public Health: Diet-related diseases are pushing our healthcare system to the brink. The rise in obesity, diabetes, and heart disease is not just a health crisis—it's an economic one that could bankrupt our nation if left unchecked.

- National Security: It might sound surprising, but the Pentagon has raised the alarm: both obesity and climate change are national defense crisis. We're running out of healthy recruits to fill military ranks, and changing weather patterns and resultant resource disruption – as well as the reactions of people impacted – are making the world less safe.
- Economic Impact: The annual cost of food waste in the U.S. is estimated at $411 billion.
- That's billions of dollars' worth of resources, labor, and energy quite literally thrown away. This impact is felt up and down the food system and impacts everyone.
- Environmental Degradation: From deforestation for agriculture to the pollution caused by food waste in landfills, our food system is putting immense pressure on our ecosystems.

The challenges are immense, but so is the opportunity for change. By addressing the inefficiencies and inequities in our food system, we have the potential to make significant strides in fighting climate change, improving public health, and creating more resilient communities.

The Path Forward

As we've seen, the problems in our food system are interconnected. That means the solutions must be too. We need approaches that address multiple issues simultaneously—reducing waste, improving access, supporting sustainable agriculture, and mitigating climate change.

Traditional charity models and top-down policy approaches, while well-intentioned, have proven insufficient to address the scale and complexity of these challenges. We need innovative, marketaligned strategies that can be effectively scaled to meet the need and create lasting change.

In the chapters that follow, we'll explore these strategies in depth. We'll look at reimagined grocery stores bringing healthy food to underserved communities, innovative food rescue operations turning would-be waste into valuable resources, and cutting-edge technologies creating more transparent and efficient food supply chains.

But more than that, we'll explore a new way of thinking about our food system—one that recognizes the connections between our plates, our communities, and our planet. One that harnesses the power of markets to serve people and the environment. One that sees food not just as a commodity, but as a tool for social and environmental justice.

This book isn't just about identifying problems; it's a call to action and a roadmap for change. We'll explore innovative, market-aligned strategies that can transform our food system from the ground up. From reimagined grocery stores to cutting-edge technology, we'll discover how we can nourish our communities, support local economies, and heal our planet - all at the same time.

The challenges we face are daunting, but the potential for positive change is enormous. By understanding the true scope and interconnected nature of our food system crisis, we can begin to craft solutions that are equally comprehensive and transformative.

As we move forward, keep this big picture in mind. The problems in our food system didn't develop overnight, and they won't be solved with quick fixes or siloed approaches. But with creativity, commitment, and a willingness to challenge the status quo, we can build a food system that nourishes our bodies, our communities, and our planet.

As we embark on this journey together, I invite you to open your mind to new possibilities. The solutions we need might not look like what we expect. They might be as simple as a bulk bin of locally-sourced grains or as complex as a blockchain-enabled food rescue network. But one thing is certain: the power to change our food system lies in our hands. Are you ready to dig in?

Take Action Now:

1. Conduct a personal food waste audit: Track your household food waste for a week. How much could you reduce?
2. Research food deserts in your area: Use the USDA's Food Access Research Atlas to identify areas of low food access in your community.
3. Calculate your food's carbon footprint: Use an online calculator to estimate the climate impact of your diet.
4. Investigate local food rescue initiatives: Are there organizations in your area working to reduce food waste? How could you support their efforts?
5. Start a conversation: Share what you've learned about the food system crisis with friends and family. Awareness is the first step towards change.

6. Research innovative food projects in your area. Are there community gardens, food co-ops, or local food hubs? How might these models be expanded or replicated?

In the marriage of market efficiency and social impact, we find the seeds of lasting change.

Chapter 2

THE CASE FOR MARKET-ALIGNED STRATEGIES

In March 1992, the streets of Los Angeles erupted in chaos. As a 19-year-old college dropout living near the University of Southern California campus, I found myself witnessing history unfold in realtime. The riots that followed the Rodney King verdict weren't just a social upheaval; they became a stark lesson in the critical importance of food access and community resilience.

As fires raged and violence spread, I learned a truth that would shape my understanding of food systems for decades to come: in times of crisis, grocery stores become lifelines. Almost every supermarket in South

Central Los Angeles had been destroyed, leaving millions without access to food and essentials. The Governor called in the National Guard to protect the 32nd Street Market, the last remaining food source in the area. My friends and I smuggled coffee and doughnuts out of the USC cafeteria to bring to the troops, and I got a front row seat to the incredible story unfoldingin front of me just a few weeks before I shipped off to basic training in the Army.

Thanks to the bravery of the National Guard and local ministers who stood vigil, that small market became a beacon of hope and a crucial pipeline for food distribution as the community struggled to recover. This experience opened my eyes to the vital role that local food access plays in community health and stability.

But the story of the 32nd Street Market didn't end there. In 2005, thirteen years after surviving the riots, it closed its doors for good when its landlord, the University of Southern California, decided not to renew its lease and razed the location to expand the school. The community it had served since 1960 became a food desert, a fate shared by countless neighborhoods across America.

This pattern - the closure of local grocery stores and the abandonment of poor communities - has been playing out across the United States for decades. It's a stark illustration of the limitations of our current food system and the urgent need for new, innovative approaches.

The Evolution and Limitations of Traditional Models

To understand how we arrived at this point, we need to look back at the origins of the modern supermarket. In the 1930s, supermarkets were developed in the United States as a marketing campaign to showcase America's agricultural might and economic strength to the world. The focus was on abundance and variety, not efficiency or equitable access.

This model worked well for its time. The sight of shelves piled high with food created a swell of support for the U.S. economic system. America may have won the Cold War thanks to the "Farms race" as much as the Arms race. But at what cost?

Today, we're seeing the long-term consequences of this approach:

1. Food Deserts and Swamps, aka Food Apartheid: As large chains abandon low-income areas, food deserts proliferate. These are often replaced by fast food outlets and corner stores selling

primarily unhealthy options, creating what the USDA calls "food swamps."

2. Inefficiency and Waste: Despite the abundance in supermarkets, 40% of the food we produce every year ends up in landfills or as animal feed. More than a quarter of that waste comes from retailers.
3. Nutritional Inequality: The foods that typically last long enough to reach poor families through food banks are often the least healthy options - sugary baked goods and candy rather than fresh produce.
4. Community Vulnerability: As we saw in Los Angeles in 1992, and again during the COVID-19 pandemic, communities without robust local food systems are extremely vulnerable to disruptions.

The Hidden Costs of Our Current System

The impacts of our failing food system go far beyond empty shelves or closed stores. They touch every aspect of our society:

1. Health Crisis: Obesity is at an all-time high and continues to rise. Diabetes, heart disease, and other preventable diet-related diseases now make up seven of the top ten killers in the U.S.

2. Economic Burden: The U.S. Center for Disease Control estimates the annual cost of chronic disease will rise to $1.1 trillion by 2050 if we don't change course.
3. Environmental Damage: Food waste in landfills contributes significantly to greenhouse gas emissions, while long-distance food transportation adds to our carbon footprint.
4. Community Resilience: The COVID-19 pandemic exposed how vulnerable our communities are to disruptions in the global supply chain.

The Need for a New Approach

It's clear that our current systems - both the profit-driven supermarket model and the charity-based food bank model - are failing to meet the needs of millions of Americans. Traditional government interventions, while important, often move too slowly to keep pace with the rapidly evolving challenges in our food system.

We need solutions that can operate at scale, that are self-sustaining, and that address multiple issues simultaneously. This is where market-aligned strategies come in.

Introducing Market-Aligned Strategies

Market-aligned strategies in food system reform harness the efficiency, innovation, and scalability of market mechanisms - but direct them towards social and environmental goals. They're about creating sustainable, scalable solutions that generate enough revenue to support and expand their impact.

Key principles of market-aligned strategies include:

1. Financial Sustainability: These initiatives generate their own revenue, reducing or eliminating reliance on grants and donations.
2. Scalability: By leveraging market mechanisms, these strategies can grow to meet the full extent of need.
3. Innovation: The competitive nature of markets drives continuous improvement and adaptation.
4. Efficiency: Market forces help allocate resources where they're needed most, reducing waste and improving overall system efficiency.
5. Stakeholder Alignment: These strategies create value for multiple stakeholders - from farmers to consumers to communities - aligning interests and fostering collaboration.

Case Studies: Market-Aligned Success Stories

To understand the potential of market-aligned strategies, let's look at some real-world examples:

1. Daily Table: Founded by the former president of Trader Joe's, this non-profit grocery store chain sells healthy food at affordable prices in underserved areas.
2. Imperfect Foods: This grocery delivery service focuses on reducing food waste by selling "ugly" produce and surplus food that would otherwise be discarded.
3. DC Central Kitchen: This organization goes beyond traditional food banks by providing culinary job training to unemployed adults while addressing food insecurity.

Overcoming Objections

Of course, the idea of using market mechanisms to address social issues isn't without its critics. Some worry that it's "just capitalism in disguise" or that it will put traditional charities out of business.

These are valid concerns, but it's crucial to understand that market-aligned strategies aren't about replacing charity or maximizing profit. They're about creating sustainable solutions that can operate at the scale needed to address systemic issues. They complement, rather than replace, traditional charity and government programs.

The Path Forward

As we face these enormous challenges in our food system, we need every tool at our disposal.

Market-aligned strategies offer a powerful approach to create lasting, scalable change.

Reflecting on my experience during the LA riots, I'm struck by how different things might have been if communities had access to resilient, locally-focused food systems. Imagine grocery stores designed not just for profit, but for community health and stability. Imagine food rescue operations that turn would-be waste into valuable resources for those in need.

This isn't just a dream - it's a viable path forward. In the chapters that follow, we'll explore specific models and strategies in depth, including an innovative approach to grocery stores that could transform food access in communities across America.

The food system revolution isn't about destroying our current system - it's about transforming it from within, using the power of markets to serve people and the planet. It's about creating a food system that works for everyone, not just a privileged few.

As we move forward, I invite you to keep an open mind, challenge your assumptions, and think creatively about how you might apply these principles in your own community. The future of our food system depends on it. [Insert "Take Action Now" box:

1. Research market-aligned food initiatives in your area. Are there any innovative grocery stores, food rescue operations, or delivery services addressing food system issues?
2. Conduct a "food audit" of your local area. Identify food deserts, existing grocery stores, and potential locations for innovative food access solutions.
3. Start a discussion in your community about food access and resilience. Share what you've learned and gather input from diverse stakeholders.
4. Support existing market-aligned initiatives. Shop at stores that prioritize local, sustainable products or use services that reduce food waste.

5. Investigate the possibility of starting a small-scale market-aligned project, like a community-supported agriculture (CSA) program or a food waste reduction service for local businesses.]

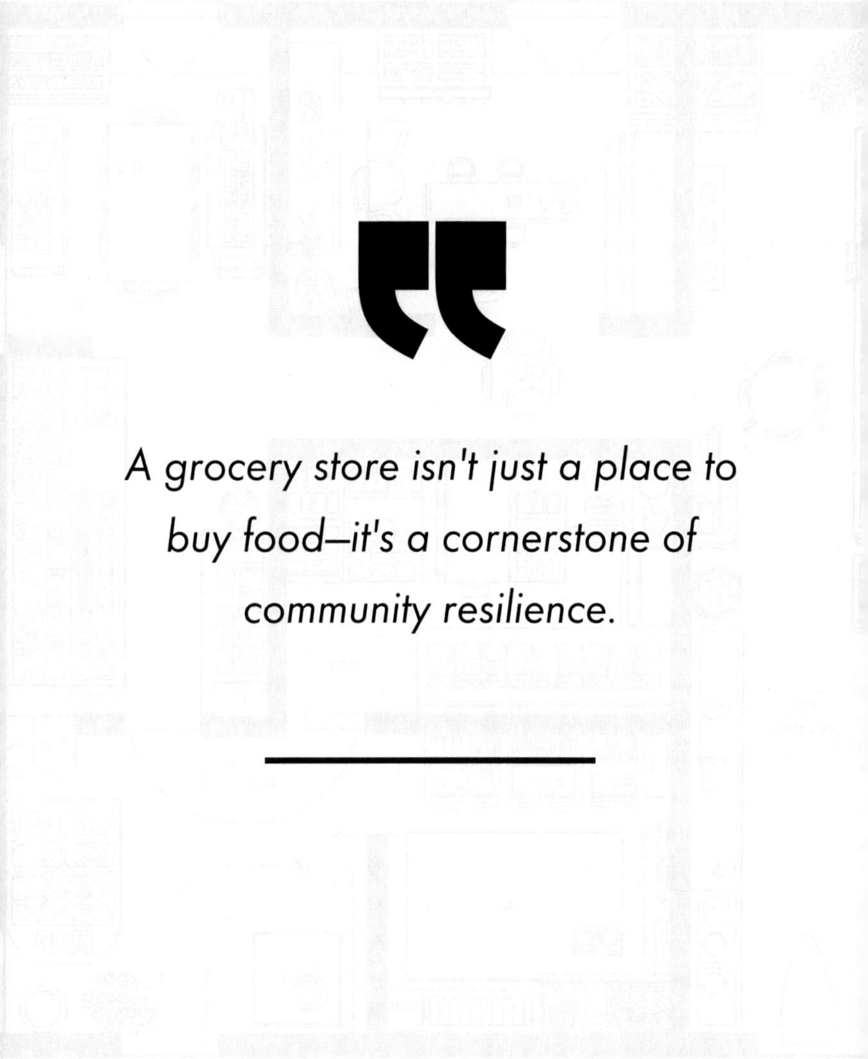

A grocery store isn't just a place to buy food—it's a cornerstone of community resilience.

Chapter 3

THE INNOVATIVE GROCERY STORE MODEL

In the heart of many American communities, a paradox unfolds daily. Supermarket shelves overflow with options, yet for millions, accessing fresh, healthy food remains a struggle. The modern supermarket, for all its abundance, has failed to serve vast swaths of our population.

I witnessed this firsthand in downtown Reno. As I watched my friends and customers at 5 Star Saloon struggle with diet-related health issues, I couldn't help but notice the glaring absence of a grocery store in the area. The mayor's public plea in 2015, "Why can't we get a grocery store into downtown Reno?" echoed in my mind, revealing a fundamental misunderstanding of the problem. It wasn't about attracting a big chain store; it was about reimagining what a grocery store could be.

Reimagining the Country Store for the 21st Century

To understand the future of grocery retail, we need to look to the past. Before the rise of supermarkets in the 1930s, most Americans shopped at small, local stores that focused on dry goods and staples. These country stores were the backbone of community food systems, providing essential ingredients without the frills and excess of modern supermarkets.

My "aha" moment came in an unlikely place—the bulk food section of a large supermarket. Suddenly, it became clear: nearly everything we need to cook healthy meals was right there, in those simple, unbranded containers. This moment of clarity would become the seed of a revolution in how we think about grocery stores and food access.

Our model takes this time-tested concept and reimagines it for the 21st century. Picture a store that fits into just 1,000 square feet, carries only what you truly need, and can thrive in both urban food deserts and rural communities. This isn't a fantasy—it's a viable model that addresses many of the shortcomings of our current food system.

Benefits of Focusing on Pantry Staples

By concentrating on dry goods and pantry staples, we can create a store that's efficient, reduces waste, and provides high-quality, nutritious ingredients at affordable prices. Here's how:

1. Reduced Food Waste: Dry goods have a longer shelf life, dramatically cutting down on spoilage. This not only reduces waste but also lowers operating costs, savings that can be passed on to customers.
2. Lower Labor Costs: With less perishable inventory, these stores require fewer staff to manage stock. This allows for better wages for employees while keeping prices low for customers.
3. Smaller Footprint: A focus on staples allows the store to operate in spaces too small for traditional supermarkets. This opens up possibilities for locating in underserved areas where large stores aren't viable.

4. Elimination of Branded Products: Removing branded items cuts out middlemen and reduces costs. It also simplifies choice for customers, reducing decision fatigue.
5. Higher Quality Ingredients: Cost savings allow for stocking better quality, often organic, options. This makes healthy eating more accessible to all income levels.

But this model isn't just about efficiency—it's about empowerment. By reintroducing communities to cooking with basic ingredients, we're providing tools for long-term food security and health.

We're not just feeding people; we're teaching them to feed themselves.

Addressing Food Deserts and Promoting Healthy Eating

Food deserts—areas where residents lack access to affordable, nutritious food—are a stark illustration of the failings of our current food retail system. These areas are disproportionately found in low-income neighborhoods and communities of color, perpetuating cycles of poor health and economic disadvantage.

Our grocery model is specifically designed to thrive in these underserved areas. Its small footprint and focus on staples make it economically viable in locations where traditional supermarkets can't or won't operate. But it goes beyond mere presence—it's about creating a hub for healthy eating and community empowerment.

Imagine a store where cooking classes are held regularly, teaching customers how to create nutritious meals from pantry staples. Picture a place where nutritionists offer free consultations, helping community members make healthier choices. Envision a space where local farmers can sell their produce, creating a direct farm-to-table pipeline in urban areas.

This isn't just about selling food—it's about transforming food cultures. By making healthy ingredients affordable and accessible, and providing the knowledge to use them, we can start to shift eating habits and improve health outcomes.

Bridging Urban and Rural Divides

One of the most exciting aspects of this model is its potential to bridge the gap between urban and rural communities. For farmers, especially those transitioning to regenerative practices, these stores provide a direct pipeline to new markets. For urban residents, they offer access to highquality, locally sourced ingredients.

In Colorado's San Luis Valley, for instance, a small store operating on this model could serve as a crucial outlet for local farmers while providing fresh, affordable food to a historically underserved area. Meanwhile, in downtown Denver, a similar store could connect city dwellers with the agricultural bounty of the state's rural areas.

This connection goes beyond just moving food—it's about building understanding and solidarity between urban and rural communities. It's about helping city dwellers understand where their food comes from and the challenges farmers face. It's about giving rural producers a stake in urban food security.

Technology and Accessibility

While the core concept of our reimagined grocery store is low-tech, we can leverage modern technology to enhance accessibility and education. In-store displays could provide recipe ideas and cooking tips, helping customers make the most of the staple ingredients. Digital scales and self-checkout systems can keep labor costs low while providing a smooth shopping experience.

One piece of technology I'm particularly excited about is the Instant Pot. These versatile cooking devices slipped below the radar of most folks as just another gadget taking up space on the counter when they came on the market in 2010, but due to their unique capacity to apply pressurized steam to rehydrate food while it cooks they can prove to be game-changers for people living in challenging environments—from converted weekly motels to dorm rooms. They allow for easy, nutritious meals using just pantry staples, even without a full kitchen. By demonstrating and potentially offering these devices, our stores can further empower communities to cook healthy meals regardless of their living situation.

Community Building Through Food

This grocery model isn't just about selling food—it's about building community. Each store can serve as a hub for nutrition education, cooking classes, and community events. Imagine a monthly "Pantry Staples Challenge," where community members come together to share meals created entirely from the store's ingredients.

These events do more than teach cooking skills—they foster connections between neighbors, create a sense of shared purpose, and help bridge the gap between food producers and consumers. They turn the act of grocery shopping from a chore into a community-building experience.

Resilience in Times of Crisis

Reflecting on my experience during the LA riots, I'm struck by how this model could enhance community resilience. A network of small, locally-focused grocery stores could provide crucial food access during emergencies, much like the 32nd Street Market did in 1992. By focusing on shelfstable staples and building strong community connections, these stores could become anchors of stability in uncertain times.

Challenges and Considerations

Of course, no model is without its challenges. Zoning laws, distribution logistics, and changing consumer habits are all hurdles we'll need to overcome. But the potential benefits—improved food access, support for local agriculture, reduced food waste, and stronger communities—make it a challenge worth tackling.

It's also important to note that this model isn't about replacing other grocery options. Rather, it's about filling a crucial gap in our food system, providing a sustainable, community-oriented alternative that can thrive where traditional supermarkets cannot.

A Call to Action

Reimagining the grocery store is more than a business proposition—it's a call to rethink our entire relationship with food. It's about connecting urban and rural communities, supporting local agriculture, reducing waste, and empowering people to take control of their nutrition.

As you finish this chapter, I challenge you to look at your own community with fresh eyes. Where are the food deserts? Where are the struggling farmers? How could a reimagined country store bridge these gaps?

The journey from my nightclub in Reno to this vision of a revolutionized food system has been long and winding. But with each step, I've become more convinced that solutions to our most pressing food issues are within reach—if we're willing to think creatively and act boldly.

Take Action Now

1. Conduct a "food audit" of your local area. Identify food deserts, existing grocery stores, and potential locations for a small-format grocery store.
2. Experiment with cooking meals using only pantry staples. Share your experiences and recipes with friends and family to demonstrate the potential of this model.
3. Research local zoning laws and regulations regarding food retail. Are there barriers to implementing a country store model in your area?
4. Start a community discussion group about food access. Invite local leaders, business owners, and residents to brainstorm solutions.
5. Explore bulk buying options in your area or online. Organize a group buy to test the waters and demonstrate the potential of this approach.]

The grocery store is just the beginning. The real revolution is yet to come. In the next chapter, we'll explore how this reimagined grocery store fits into a larger system—the nexus—that has the potential to transform not just how we shop for food, but how we grow it, distribute it, and think about it.

Every Node is a seed of change, every Nexus a garden of possibility. Together, they grow a forest of resilience.

Chapter 4

THE HUB AND SPOKE REVOLUTION: NEXUS AND NODES

"

Introduction: Beyond the Grocery Store

In the previous chapter, we explored how reimagining the grocery store can transform food access in underserved communities. But this innovative model is just one piece of a larger, more revolutionary system. Welcome to the world of Nexus and Nodes - a comprehensive approach to reshaping our entire food ecosystem.

The Nexus Concept

The Nexus is the heartbeat of our reimagined food system. It's not just another warehouse or distribution center - it's a multifaceted powerhouse of food rescue, processing, community empowerment, and sustainable agriculture.

My journey to developing this concept was shaped by my diverse experiences - from witnessing the critical role of food access during the LA riots, to my time in the Army where I honed my skills in logistics and coalition-building, to my years as a small business owner. Each of these experiences contributed to my understanding of how a truly effective food system needs to operate.

The Nexus addresses multiple issues simultaneously: food waste, access to healthy food, community resilience, and environmental sustainability. It's designed to be a force multiplier, amplifying the impact of existing community efforts while catalyzing new initiatives.

Key Components of the Nexus

a. Food Processing Hub

At the core of the Nexus is a large-scale commercial kitchen equipped with state-of-the-art technology. This isn't just about reheating meals - it's about transforming would-be waste into valuable, nutritious food products.

One of our game-changing technologies is the blast chiller. These industrial-scale freezers can rapidly cool food from cooking temperatures to freezing in a matter of minutes. This not only allows us to rescue and preserve large quantities of prepared foods that would otherwise go to waste, but it also dramatically extends the shelf life of fresh produce and prepared meals.

But we're not stopping at preservation. Our processing capabilities allow us to create shelf-stable foods, forming a direct link back to our node grocery stores. Imagine turning excess produce into canned goods, sauces, or dehydrated ingredients - all of which can be sold as pantry staples in our nodes and other retail outlets.

This processing power also opens up exciting possibilities for developing our own food brands. By utilizing manufacturing-scale equipment, we can produce high-quality, affordable food products that not only stock our own shelves but could potentially be marketed to other retailers as well. This creates another revenue stream to support our mission while providing communities with healthy, accessible food options. b. Distribution Center

The Nexus serves as the central hub in our hub-and-spoke distribution model. From here, we coordinate the flow of food to our network of nodes, ensuring that each community gets what it needs, when it needs it. This model dramatically reduces food miles, cutting down on transportation costs and environmental impact. c. Community Spaces

The Nexus isn't hidden away in an industrial park. It's a vibrant part of the community, with spaces for education, events, and community gatherings. We've designed the Nexus to include a largecapacity event space, perfect for hosting everything from cooking classes to community meetings to large-scale fundraisers.

Imagine a public café where people can taste dishes made from rescued food, or an event space where community groups can host meetings and workshops. These spaces create opportunities for producers and consumers to connect directly, fostering a deeper understanding of our food system.

d. Sustainability Initiatives

Sustainability is at the heart of the Nexus design. One of our most exciting features is the anaerobic biodigester. This technology turns food waste into energy and high-quality compost, closing the loop in our food system. Not only does this reduce methane emissions from landfills, but it also provides a renewable energy source for our operations and valuable fertilizer for local farmers.

We're also incorporating solar panels to power our operations and implementing water recycling systems to minimize our environmental impact. e. Agricultural Component

The Nexus isn't just about processing and distributing food - it's also about growing it. We've incorporated an agricultural component that serves both educational and production purposes. This could include demonstration gardens, hydroponics systems, and even urban farming initiatives.

One innovative aspect is our plan to use shipping containers for fresh food production. These can be implemented at both the Nexus and node levels, allowing us to grow fresh produce with minimal food miles. Imagine picking a salad for your dinner from a garden right outside your local grocery store!

This agricultural element completes our closed-loop system. We're not just distributing food - we're growing it, processing it, selling it, and even turning the waste back into energy and fertilizer to grow more food. It's a scalable model that can be adapted to meet the needs of various communities.

The Nodes: Reimagined Grocery Stores

Our nodes - the reimagined grocery stores we explored in the previous chapter - are the points where our system directly touches individual lives. Each node is customized to meet the unique needs of its community, stocking culturally appropriate foods and offering programs that resonate with local needs.

While the Nexus provides large-scale event spaces, the nodes rely more on partnerships with existing community assets like parks and community centers. This allows us to keep the footprint of each node as compact as possible while still providing vital community services.

Innovative Technologies as Force Multipliers

While the core of our model is about reconnecting communities with their food systems, we're not shy about embracing cutting-edge technologies to amplify our impact.

Blockchain technology, for instance, offers exciting possibilities for food tracking and waste reduction. Imagine being able to trace a tomato from the moment it's picked to when it lands on your plate. This level of transparency not only helps us minimize waste but also builds trust with consumers.

Here's how it could work: When a restaurant donates excess meals, we can use blockchain to assign a value to that donation. If the food is processed and resold, we can provide a portion of the profits back to the donor, creating a financial incentive for participation in our food rescue program. This same system can track temperatures for food safety, donation destinations, and even inputs and outputs for our biodigester, creating a fair credit system for compost pricing.

Artificial Intelligence is another powerful tool in our arsenal. By analyzing patterns in food production, consumption, and waste, AI can help us predict needs and optimize our operations. This could mean anything from adjusting our food rescue routes for maximum efficiency to predicting which crops will be in surplus in the coming months, allowing us to plan our processing accordingly.

These technologies aren't just flashy add-ons - they're integral to creating a food system that's efficient, transparent, and responsive to community needs.

Economic and Social Impact

The Nexus and Node model isn't just about moving food around - it's about creating a new economic engine for communities.

Job Creation and Skill Development: From kitchen staff processing rescued food to logistics experts managing our distribution, to retail workers in our nodes, we're creating a range of jobs that can't be outsourced. More than that, we're building careers. Through our training programs, we're offering people the chance to gain valuable skills in food preparation, sustainable agriculture, logistics, and retail management.

Supporting Local Farmers and Producers: Our model provides a reliable market for local farmers, especially those transitioning to regenerative practices. By connecting them directly with urban consumers, we're helping to build a more resilient local food economy.

Building Community Resilience: In an era of increasing uncertainty - from economic shocks to natural disasters - our model builds resilience from the ground up. Our nodes can serve as emergency food distribution centers in times of crisis, while our Nexus can quickly pivot to support community needs.

Challenges and Solutions

Of course, a model this ambitious comes with its share of challenges. Zoning laws and regulations designed for traditional food retail and distribution models may need to be updated. We're working closely with local officials to help them understand the benefits of our approach and to find creative solutions within existing frameworks.

Funding is another key challenge. Traditional funding models, whether charitable or commercial, aren't always well-suited to our hybrid approach. That's why we're exploring innovative funding mechanisms, including impact investing and community ownership models. We'll dive deeper into these in the next chapter.

Changing consumer habits is perhaps our biggest hurdle. We're asking people to rethink their relationship with food, from how they shop to how they cook. But we've found that when people understand the impact of their choices - on their health, their community, and the planet - they're eager to be part of the solution.

Vision for the Future

Imagine a network of these systems spread across the country, each adapted to its local context but connected in a web of mutual support and shared learning. We could see:

- Dramatic reductions in food waste and associated greenhouse gas emissions
- A revitalization of rural economies and a new generation of sustainable farmers
- Significant improvements in public health and reductions in diet-related diseases
- More resilient local food systems better able to withstand shocks and disruptions
- A transformation in how people relate to their food, their communities, and the environment This isn't just a dream - it's a viable path forward. But it will take all of us to make it happen.

Call to Action

As we close this chapter, I invite you to think about how this model could transform your own community. What would a Nexus look like in your area? What local challenges could it address? What local strengths could it amplify?

The food system revolution isn't something that's happening somewhere else, led by someone else. It's happening right here, right now, and it's led by people like you. So roll up your sleeves, connect with your community, and let's get to work. The future of our food system - and our planet - depends on it.

[Insert "Take Action Now" box:

1. Research potential locations for a Nexus in your region. Consider factors like proximity to urban centers, agricultural areas, and transportation networks.
2. Organize a community meeting to discuss the potential for a Nexus-style facility in your area. Invite farmers, food business owners, and community leaders.
3. Investigate underutilized commercial kitchen spaces in your community. Could these be repurposed for food rescue and processing?
4. Start a conversation with local farmers about their challenges in accessing markets. How could a Nexus and Nodes model help address these?
5. Look into local regulations around food processing and distribution. What changes might be needed to implement this model in your area?]

"

In the intersection of profit and purpose, we find the fuel for systemic change.

Chapter 5

CHANGING THE RULES: A NEW WAY TO FUND FOOD RESCUE

Introduction: The Funding Challenge

As we've explored the revolutionary potential of our Nexus and Node model, you might be wondering: "This sounds great, but how do we pay for it?" It's a fair question, and one that gets to the heart of why so many promising solutions to our food system crisis never make it off the drawing board.

The hard truth is this: our current funding systems for social impact are fundamentally broken. They're not just inefficient - they're actively stifling innovation and perpetuating the very problems they claim to solve.

The Limits of Traditional Funding

Let me paint you a picture from my own experience. After developing our innovative model for addressing food insecurity, I eagerly sought out grant funding to bring it to life. What I found was a system seemingly designed to maintain the status quo rather than drive real change.

Automated restrictions on many grant applications meant I couldn't even get my foot in the door. One major health foundation required a minimum of $250,000 in annual revenue just to apply - a catch-22 for any new initiative trying to get off the ground. Another prominent foundation doesn't accept unsolicited proposals at all, meaning unless you're already in their inner circle, your ideas - no matter how groundbreaking - will never see the light of day.

It's a system that favors large, established organizations and perpetuates a cycle where the same approaches are funded year after year, regardless of their efficacy. Meanwhile, innovative solutions from smaller organizations or individuals outside the nonprofit industrial complex are left to wither on the vine.

This isn't democratic, it isn't effective, and it certainly isn't driving the kind of systemic change we desperately need. We can and must do better.

A Global Vision for Local Impact

Here's where we flip the script. Instead of begging for scraps from a broken system, what if we could attract global investment to drive local change? What if we could leverage the power of cryptocurrency and blockchain technology to democratize funding for social impact?

Imagine a world where someone in Tokyo could invest in a food rescue hub in Denver, tracking their impact in real-time through blockchain technology. Picture a farmer in rural Colorado receiving a micro-loan funded by hundreds of small investors from around the globe, all passionate about supporting sustainable agriculture.

This isn't just a pipe dream - it's a viable path forward that aligns perfectly with our Nexus and Node model. Just as our food system model works to connect local communities with global resources, our funding model can do the same.

And here's the kicker: this model works everywhere. While my heart is set on building our pilot program here in Colorado, I'm prepared to take this vision wherever it's welcomed. If forwardthinking investors in Europe, South America, or Asia see the potential before we do here in the States, I'm ready to move. This isn't a threat

- it's an invitation to be part of a global revolution in how we fund and implement social change.

The Challenge to America

Let's be clear: Americans often fear loss more than they desire change. But in this case, the loss we should fear is the opportunity to lead in solving one of the most pressing issues of our time. Our military has identified food insecurity as an existential threat. The health and economic costs of our broken food system are staggering. Can we really afford to let other countries take the lead in solving these issues?

This is a challenge to America: step up or step aside. The food system revolution is coming, with or without us. The question is, will we lead it or follow?

A New Paradigm for Funding

So, how do we make this vision a reality? By combining innovative funding mechanisms with the kind of resourcefulness I learned running bars and nightclubs. Here's how:

1. Cryptocurrency and Blockchain: We'll use tokenization to allow fractional investment in our real estate and operations. This opens up

investment opportunities to a global pool of funders and allows for unprecedented transparency in impact tracking.

2. Multiple Revenue Streams: Just as a successful bar doesn't rely solely on alcohol sales, our Nexus and Node model incorporates diverse income sources. From food sales to processing services, event space rentals to educational programs, we're creating a resilient financial model that can weather economic fluctuations.
3. Partnerships to Reduce Costs: We'll work with municipalities to repurpose underutilized spaces, collaborate with private groups to share resources, and leverage community assets to keep our overhead low.
4. Community Ownership: Our model includes a path to community ownership, ensuring that the long-term benefits flow back to the people we serve.
5. Global Network, Local Impact: By creating a network of Nexus and Node systems worldwide, we can share best practices, leverage economies of scale, and create a powerful movement for change.

The Path Forward

This approach isn't just about funding a food rescue operation - it's about revolutionizing how we fund and implement social impact initiatives across the board. It's about creating a model that's as innovative in its financing as it is in its operations.

Are there challenges? Absolutely. Regulatory hurdles, technological learning curves, and deeply ingrained mindsets all stand in our way. But the potential benefits - a transformed food system, revitalized communities, and a new paradigm for funding social good - make it a challenge worth tackling.

The question isn't whether we can change the world. The innovators we've discussed have already proven that we can. The question is: will you join us? Will you add your voice, your hands, your heart - and yes, your investment - to this movement?

Our communities are hungry - not just for food, but for connection, for purpose, for hope. The table is set. The work is waiting. And the time is now. Are you ready to change the rules of the game?

Take Action Now:

1. Educate yourself on cryptocurrency and blockchain technology, focusing on their potential for social impact projects. Free online courses are available on platforms like Coursera or edX.
2. Organize a community meeting or online forum to discuss innovative funding models for local food initiatives. Invite local investors, business leaders, and community organizers.
3. If you're an accredited investor, explore how you might incorporate impact investing into your portfolio, particularly in food system projects.
4. Start a local investment club focused on funding community food projects. This can be a way to pool resources and knowledge while supporting local initiatives.
5. Research and connect with global organizations working on innovative funding for food systems, such as the Food Systems Finance Action Network (FSFAN) or the Global Alliance for the Future of Food.
6. If you're involved with a local food initiative, begin exploring how you might incorporate multiple revenue streams into your model to increase sustainability and impact.

The path to a sustainable future is paved with the choices we make about our food system today.

Chapter 6

BEYOND FOOD ACCESS: ENVIRONMENTAL AND SOCIAL IMPACT

“

The Ripple Effect: A Butterfly's Wings

Imagine a butterfly flapping its wings in Brazil, setting off a chain of events that culminates in a tornado in Texas. This concept, known as the butterfly effect, illustrates how small changes can have far-reaching consequences. Now, picture our Nexus and Node model as that butterfly, its wings setting in motion a cascade of positive changes that ripple out to touch every aspect of our communities and our planet.

When we first set out to address food insecurity, we couldn't have imagined the breadth of impact this model would have. But as we've developed and refined our approach, we've come to understand that food isn't just about nutrition - it's the thread that weaves together the fabric of our communities, our economies, and our environment.

A Synergistic Solution

At the core of our model lies a set of interconnected components that work together to create a powerful, regenerative system. Let's dive deep into each of these elements to understand how they synergize to create impact far beyond just feeding people.

1. Food Rescue: The Foundation of Change

Our food rescue efforts are about more than just redirecting excess food. By diverting food waste from landfills, we're preventing methane emissions, a potent greenhouse gas that contributes significantly to climate change. But the impact goes further. Every pound of food rescued represents resources conserved - water, energy, and labor that went into producing that food aren't wasted.

2. Biodigester Technology: Waste to Resource Alchemy

At the heart of our Nexus model lies a powerful piece of technology that exemplifies our commitment to closed-loop systems: the anaerobic biodigester. This isn't just a fancy composter – it's a game-changer in how we think about food waste and energy production.

How it Works: Anaerobic biodigesters use microorganisms to break down organic matter in the absence of oxygen. This process, known as anaerobic digestion, produces two valuable outputs:

- Biogas: A mixture primarily of methane and carbon dioxide, which can be used as a renewable energy source.
- Digestate: A nutrient-rich material that can be used as a high-quality fertilizer.

Our industrial-scale biodigester can process ***hundreds of tons*** of organic waste per hour, turning what would have been a liability into a valuable resource. This goes far beyond merely food waste – every scrap of animal manure, field clippings, in fact nearly any biomass can be converted. We're specifically focused on non-human waste, so unlike a municipal wastewater treatment facility our compost can be used on our food crops to amend the soil regeneratively.

Environmental Impact: The environmental benefits of this technology are substantial:

- Methane Capture: By capturing methane that would otherwise be released into the atmosphere from decomposing food in landfills, we're tackling one of the most potent greenhouse gases head-on.
- Reduced Reliance on Fossil Fuels: The biogas produced can be used to generate electricity or heat for our operations, reducing our reliance on fossil fuels.
- Soil Health: The digestate produced is a powerful soil amendment, reducing the need for chemical fertilizers and improving soil health.
- Water Conservation: The liquid portion of the digestate can be used for irrigation, helping to conserve water resources.

Economic and Community Benefits: The biodigester isn't just good for the environment – it creates economic opportunities and community benefits:

- Energy Cost Savings and Revenue Generation: By producing our own energy and selling excess biogas and high-quality compost, we create a sustainable financial model.

- Job Creation: Operating and maintaining the biodigester creates skilled job opportunities.
- Educational Opportunities: The biodigester serves as a powerful educational tool, demonstrating scalable principles of circular economy and waste-to-resource conversion.

3. Reduced Food Miles: Localizing Our Food System

Our hub-and-spoke model dramatically cuts down on transportation, reducing carbon emissions and bringing fresh food closer to consumers. By localizing our food system, we're not just reducing environmental impact - we're creating more resilient communities less dependent on long, fragile supply chains.

4. Efficiency in Distribution: One-Way Trip for Food

We're not just moving food around - we're ensuring that food enters communities once and stays there, eliminating redundant transportation and handling. This efficiency doesn't just save fuel and reduce emissions - it ensures fresher food reaches consumers and reduces the chance of spoilage along the way.

5. Improved Health Metrics: Nourishing Bodies and Communities

Access to fresh, nutritious food leads to better health outcomes, reducing the burden on healthcare systems and improving quality of life. But the health benefits extend beyond individual nutrition. By creating spaces for community gathering and education around food, we're fostering mental and social well-being too.

6. Support for Regenerative Agriculture: Healing the Land

By providing a reliable market for sustainably grown produce, we're incentivizing farmers to adopt practices that heal the soil and increase biodiversity. This shift towards regenerative agriculture has profound implications for climate change mitigation, water conservation, and long-term food security.

Healing the Urban-Rural Divide

One of the most exciting aspects of our model is its potential to bridge the gap between urban and rural communities. For too long, these two worlds have been disconnected, leading to misunderstandings and missed opportunities.

By creating direct links between urban consumers and rural producers, we're not just moving food - we're building understanding. Urban dwellers gain a deeper appreciation for where their food comes from and the challenges farmers face. Rural communities gain reliable markets for their produce and a stake in urban food security.

This reconnection has the potential to heal deep societal divides. When we break bread together, we begin to see our common humanity. When we understand each other's challenges, we can work together to find solutions.

Food as a Tool for Planetary Healing

As we face the existential threat of climate change, it's easy to feel overwhelmed. But our food system offers a powerful lever for change. By transforming how we grow, distribute, and consume food, we can make significant strides in healing our planet.

Our model addresses climate change from multiple angles:

- Methane capture through biodigesters
- Carbon sequestration through support for regenerative agriculture

- Reduced transportation emissions through efficient, localized distribution
- Decreased demand for resource-intensive animal products through a focus on plant-based staples and rescued food

Taken together, these initiatives represent a holistic approach to climate action, all centered around something as fundamental as how we feed ourselves.

Building Community Resilience

In an era of increasing uncertainty - from economic shocks to natural disasters - community resilience is more important than ever. Our model builds this resilience from the ground up:

- Strengthened local food systems reduce vulnerability to global supply chain disruptions
- Nodes serve as emergency food distribution centers in times of crisis
- Skill development in cooking and small-scale agriculture empowers individuals
- Community connections fostered through our initiatives create vital social capital

A Seat at the Table for Everyone

At its heart, our model is about more than just distributing food - it's about creating a table long enough for everyone to have a seat. This means:

- Providing access to nutritious food for all, regardless of income or location
- Creating spaces for cultural exchange and celebration of diverse food traditions
- Empowering communities to shape their own food systems
- Ensuring all voices are heard in decisions about food policy and distribution

By centering equity and inclusion in our approach, we're not just addressing food insecurity - we're building a more just and cohesive society.

Measuring Our Impact

We're committed to rigorously tracking and communicating our outcomes, developing metrics that go beyond simple pounds of food distributed or meals served:

- Reductions in greenhouse gas emissions from waste diversion and local sourcing

- Improvements in community health indicators
- Economic impacts like job creation and local business growth
- Changes in attitudes and behaviors around food and sustainability

By transparently sharing these outcomes, we hope to inspire others and contribute to a growing body of knowledge about effective food system interventions.

A Vision for the Future

Imagine if this model were adopted widely. We could see:

- A dramatic reduction in food waste and associated greenhouse gas emissions
- A revitalization of rural economies and a new generation of sustainable farmers
- Significant improvements in public health and reductions in diet-related diseases
- More resilient local food systems better able to withstand shocks and disruptions
- A transformation in how people relate to their food, their communities, and the environment This isn't just a dream - it's a viable path forward. But it will take all of us to make it happen.

The Butterfly's Journey

As we close this chapter, I invite you to think of yourself as that butterfly. The choices you make about food - where you buy it, how you prepare it, who you share it with - may seem small. But they have the power to set in motion changes that ripple out far beyond your immediate circle.

Whether you're a consumer, a farmer, a policymaker, or an entrepreneur, you have a role to play in this food system revolution. By working together, we can create a food system that nourishes our bodies, our communities, and our planet.

Are you ready to spread your wings?

Take Action Now:

1. Start a community garden or join an existing one. This helps reduce food miles and connects you directly with your food source.
2. Organize a "zero waste" community meal using only locally-sourced ingredients. This can showcase the potential of a closed-loop food system.
3. Research and support local farmers who are practicing regenerative agriculture. Consider joining a CSA (Community Supported Agriculture) program.

4. Conduct a personal or household food waste audit. Track your waste for a week and brainstorm ways to reduce it.
5. Advocate for policies that support sustainable food systems in your local government. Attend city council meetings or write to your representatives.
6. Start a neighborhood composting program to divert food waste from landfills and create valuable soil amendments.

Coalition building isn't just about aligning goals—it's about weaving together the diverse strengths of a community.

Chapter 7

BUILDING A GRASSROOTS COALITION

Introduction: The Power of Collective Action

In the world of special operations, we had a saying: "Lone wolves die alone." This principle holds true not just in military operations, but in any endeavor aimed at creating systemic change. As I've learned throughout my journey - from military service to nightclub ownership to food system reform - the key to lasting impact lies in our ability to build and nurture coalitions.

My Journey in Coalition Building

My first lessons in coalition building came in an unlikely place: the U.S. Army's Psychological Operations (PSYOP) training. While many might associate PSYOP with manipulation, at its core, it's about understanding human dynamics and building trust to achieve shared goals. We called it "winning hearts and minds," but really, it was about creating coalitions across cultural and ideological divides.

One of my proudest achievements in the Army came after an injury led to my reassignment as the armorer in my unit's supply room. I saw an opportunity to build a coalition between armorers and supply rooms throughout the special ops community at Fort Bragg. By fostering collaboration and resource-sharing, we dramatically improved efficiency and readiness. This initiative, which earned me an Army Achievement Medal, taught me the power of turning potential competitors into collaborators.

Years later, as a bar owner in Reno, I applied these same principles to transform potential competition into cooperation. By spearheading initiatives like "beer crawls" that encouraged patrons to visit multiple establishments, we turned downtown Reno into a walking destination. This co-marketing approach grew the pie for everyone, demonstrating that when we work together, we all benefit. It was this collaborative spirit that helped my bar become the top-rated dance club in northern Nevada.

These experiences - from military operations to small business entrepreneurship - have profoundly shaped my approach to building coalitions for food system reform. They've taught me that true change happens when we bring diverse stakeholders together, align interests, and work towards a common goal.

Identifying Key Stakeholders

The first step in building an effective coalition is identifying who needs to be at the table. In the context of food system reform, this goes far beyond the obvious players like farmers and food banks. We need to think broadly and creatively about who has a stake in a healthier, more equitable food system.

Consider this: When I was planning those beer crawls in Reno, I didn't just reach out to other bar owners. I engaged local law enforcement to ensure safety, city officials for permits, local media for promotion, and even nearby restaurants and hotels who would benefit from increased foot traffic.

Similarly, in food system reform, we need to engage:

- Farmers and food producers
- Food retailers and distributors
- Local government officials
- Healthcare providers
- Educators and academic institutions
- Environmental organizations
- Community leaders and activists
- Local businesses
- Technology innovators

Remember, unlikely allies can often be your most powerful partners. Just as I found common ground between armorers and supply officers in the Army, we can find shared interests between seemingly disparate groups in the food system.

Strategies for Engagement

Once you've identified your stakeholders, the next challenge is engaging them effectively. Here are some strategies I've found successful:

1. Speak Their Language: In PSYOP training, we learned the importance of tailoring our message to our audience. When talking to a farmer, focus on how your initiative can provide stable

markets. When engaging healthcare providers, emphasize the potential health benefits.

2. Find Common Ground: Even groups that seem at odds often have shared interests. Your job is to identify and emphasize these commonalities.
3. Create Win-Win Scenarios: Like the beer crawls that benefited all participating businesses, look for ways your food system initiatives can create value for all stakeholders.
4. Build Personal Relationships: Coalition building is fundamentally about human connections. Take the time to understand each stakeholder's needs, fears, and aspirations.

Overcoming Silos

One of the biggest challenges in food system reform is breaking down the silos that often exist between different sectors. In the Army, I saw firsthand how these silos could impede effectiveness. The solution? Creating opportunities for cross-functional collaboration and communication.

Consider organizing regular "food system roundtables" where diverse stakeholders can meet, share perspectives, and collaborate on solutions. These gatherings can be the genesis of innovative partnerships and ideas.

Youth Engagement and Empowerment

In my experience, young people are often the most passionate and innovative change-makers. They bring fresh perspectives and energy that can revitalize any coalition. In Reno, some of our most successful events were driven by ideas from young staff members.

Consider creating a "Youth Food Council" as part of your coalition. Give young people real responsibility and decision-making power. Their insights might surprise you, and their energy will certainly inspire you.

Leveraging Technology for Coalition Building

While face-to-face relationships are crucial, technology can be a powerful tool for coordination and impact tracking. In the Army, we used advanced systems to coordinate complex operations. In the food system, we can use similar tools to track food flows, measure impact, and coordinate coalition activities.

Consider implementing a shared digital platform for your coalition. This could include tools for communication, resource sharing, and impact measurement.

Navigating Challenges

Coalition building isn't always smooth sailing. Conflicts will arise, momentum will flag, and obstacles will emerge. The key is to anticipate these challenges and have strategies in place to address them.

One approach I've found effective is to create a "coalition charter" that outlines shared goals, decision-making processes, and conflict resolution procedures. Having these agreements in place can help navigate difficult moments.

The Future of Food System Coalitions

As we look to the future, I see tremendous potential for food system coalitions to drive transformative change. Imagine a global network of local food coalitions, sharing best practices, resources, and support. This isn't just a dream - it's a viable path forward, and one that I believe is crucial for addressing the complex challenges we face.

Call to Action: Join the Movement

Building effective coalitions isn't easy, but it's essential for creating lasting change. Whether you're a farmer, a policymaker, a business owner, or simply someone who cares about creating a better food system, you have a role to play in this movement.

Remember, every great coalition starts with a single connection. Who in your community could you reach out to today to start building a coalition for food system change? Take Action Now: [BOX]

1. Identify 3-5 key stakeholders in your local food system and reach out to them to start a conversation.
2. Organize a "food system roundtable" in your community, inviting diverse stakeholders to discuss local challenges and opportunities.
3. Start a "Youth Food Council" or support an existing one in your area.
4. Research digital tools that could help coordinate food system activities in your community.
5. Draft a "coalition charter" for a local food system initiative, outlining shared goals and processes.
6. Attend a local government meeting and speak up about the importance of collaborative food system reform.

Remember, in coalition building, as in special ops, we're stronger together. Let's build a movement that can transform our food system from the ground up.

Real growth isn't measured in dollars or square footage, but in lives improved and communities strengthened.

Chapter 8

REDEFINING GROWTH: A NEW PARADIGM FOR FOOD SYSTEM IMPACT

"

Introduction: Beyond Traditional Expansion

When I transitioned from running one of Reno's most successful nightclubs to reimagining our food system, I brought with me a fundamental truth: real growth isn't just about getting bigger—it's about creating deeper, more meaningful impact.

At 5 Star Saloon, we didn't just serve drinks; we built a home. It was an LGBT icon, a community center with a great vodka selection and a killer dance floor. Our success wasn't measured in profits alone, but in the strength of the community we fostered. This approach left our competitors scratching their heads, unable to replicate what we had created.

Now, as we tackle the challenges of our food system, it's time to get those heads scratching again. We need to redefine what growth and success look like in this critical sector.

Reframing the Concept of Growth

In the world of food systems, traditional notions of growth—more stores, more revenue, more market share—fall woefully short. Instead, we need to conceptualize growth as increasing our ability to serve needs and even identify new ones over time.

For instance, discovering that the Denver school system needs support with light add-value processing for school lunches, or realizing we can share our contact info with truck drivers so they have somewhere to take food if a store rejects a delivery—these are examples of growth. They represent an expansion of our impact and our ability to respond to community needs.

This responsive, needs-based approach to growth allows us to remain flexible and adaptive. Instead of creating rigid standard operating procedures that are difficult to pivot from if circumstances change, we position ourselves to be responsive to evolving community needs.

Principles Guiding Sustainable, Impactful Growth

Our approach to growth is guided by several key principles:

1. Climate Focus: By centering our efforts on combating climate change, we align ourselves with a critical global need and open up opportunities for measurable impact.
2. Integrity: We stand by our commitments and maintain transparency in all our operations.
3. Solidarity: We stand with and for the communities we serve, amplifying marginalized voices.
4. Courage: We're not afraid to challenge the status quo when it's not serving community needs.
5. Steadfastness: We remain committed to our mission, even when faced with obstacles.
6. Conscientiousness: We consider the broader implications of our actions on communities and the environment.

7. Honor: We uphold our promises and maintain high ethical standards in all our dealings.
8. Agile Responsiveness: We cultivate a culture of flexibility and swift action, ready to pivot our strategies in response to emerging community needs, unforeseen challenges, or opportunities for amplified impact.

Community Ownership and Empowerment

Perhaps our most revolutionary idea is leveraging private equity to develop the solution, and then selling the developed business and property to the community as a cooperative model. This approach, baked into our plan from the start, is a concrete way to share equity with the community—something that rarely happens in traditional business models.

This isn't just theory—it's an extension of how we operated at 5 Star Saloon. We didn't just serve our community; we made them co-creators of the experience. From involving staff in decision-making to encouraging regulars to shape our events, we created a sense of ownership that went far beyond the four walls of the bar.

In our food system model, this translates to:

1. Transparent Decision-Making: Involving community members in key decisions about store offerings, programming, and more.
2. Skills Development: Offering training and education programs that empower community members to take on leadership roles.
3. Profit-Sharing: Structuring our model so that profits flow back into the community.
4. Gradual Transition: Planning for a phased handover of ownership and control to the community over time.

Multiple Revenue Streams for Sustainability

Just as a successful bar doesn't rely solely on alcohol sales, our Nexus and Node model incorporates diverse income sources. At 5 Star, we charged cover, rented space for gaming machines and pool tables, hosted private events, sold food, and more. Each revenue stream not only contributed to our bottom line but also enhanced the overall experience for our patrons.

In our food system model, this translates to:

1. Food Sales: Both retail and wholesale, including value-added products created from rescued food.
2. Processing Services: Offering co-packing and processing services to local producers.
3. Education Programs: Cooking classes, nutrition workshops, and more.
4. Event Space Rentals: For community gatherings, private events, and partner organizations.
5. Waste Management Services: Using our biodigester to process organic waste for other businesses.

This diversity of revenue streams not only ensures financial sustainability but also allows us to serve our community in multiple ways.

Building Resilience Through Redundancy

In the military, redundancy is key—we don't go into battle with barely enough food or weaponry; we go in expecting failures and build that into every battle plan. This principle applies equally to our food system model.

We're not just creating a single point of food access; we're building a resilient network of interconnected nodes, each capable of supporting the others. If one node faces challenges, the others can step in to ensure continuous service to the community.

This redundancy extends to our funding model, our supply chains, and our leadership structure. By building multiple fail-safes into our system, we ensure that our mission to provide equitable food access can continue even in the face of unexpected challenges.

Measuring Growth and Impact

Traditional metrics like revenue and profit margins don't capture the full impact of our model. Instead, we focus on metrics that truly matter to communities:

1. Food Security: Measuring improvements in access to healthy, affordable food.
2. Health Outcomes: Tracking changes in diet-related health issues in our communities.
3. Environmental Impact: Quantifying reductions in food waste and carbon emissions.
4. Economic Empowerment: Measuring job creation and local economic growth.
5. Community Engagement: Tracking participation in programs, events, and decision-making processes.

By focusing on these metrics, we ensure that our growth is always aligned with our mission of creating healthier, more resilient communities.

A New Vision for Growth

Imagine a food system where growth isn't measured by how many stores we open, but by how many communities we empower. Where success isn't about market domination, but about market creation—building new, equitable economies around food.

This isn't just a dream—it's a viable path forward. But it requires us to be bold, to challenge conventional wisdom, and to put communities at the center of everything we do.

As we move forward with this model, we're not just changing how food is distributed—we're changing the very definition of what a successful food business looks like. And in doing so, we're creating a blueprint for how other industries can redefine growth to prioritize community and environmental well-being.

Call to Action: Be a Catalyst for Change

The food system revolution isn't something that's happening somewhere else, led by someone else. It's happening right here, right now, and it's led by people like you. Whether you're a community leader, an entrepreneur, a policymaker, or simply someone who cares about these issues, you have a role to play in redefining what growth and success look like in our food system.

So I challenge you: Look at your community with fresh eyes. Where are the opportunities to create food businesses that truly serve community needs? How can you incorporate principles of community ownership and empowerment into your work?

Remember, every great movement starts with individuals who dare to imagine a different world. The seeds of a revolution have been planted. With this book, you hold the blueprint for nurturing those seeds into a flourishing, resilient food system that can weather any storm.

Take Action Now:

1. Conduct a community food system assessment. Identify gaps in service and opportunities for community-centered growth.
2. Start a discussion group about alternative business models that prioritize community ownership and empowerment.

3. Research cooperative ownership models and how they might be applied to food businesses in your area.
4. Develop a set of community-centered metrics to measure the success of local food initiatives.
5. Organize a workshop on multiple revenue stream strategies for local food businesses.
6. Reach out to local policymakers about incentives for community-owned food businesses.

The table is set. The work is waiting. And the time is now. Are you ready to redefine growth and create a food system that truly serves all?

Success in transforming our food system isn't a destination—it's a journey of constant learning, adapting, and growing together.

Chapter 9

MEASURING SUCCESS AND CONTINUOUS IMPROVEMENT - CULTIVATING A REVOLUTION IN FOOD SYSTEMS

As we near the end of our journey through this reimagined food system, it's time to address a critical question: How do we know we're succeeding? More importantly, how do we ensure that we continue to improve and adapt in a world of constant change?

In my years of experience - from military service to running nightclubs to reimagining food systems - I've learned that success is never a destination. It's a continuous journey of learning, adapting, and growing.

Let's explore how we can apply this mindset to our food system revolution.

Metrics That Matter: Painting the Picture of Progress

Traditional measures of success in the food industry often focus on profit margins, market share, and expansion rates. But in our model, success looks different. We're not just building businesses; we're cultivating healthier communities, a more resilient planet, and a more equitable society.

Here are some key areas we focus on when measuring our impact:

1. Food Security and Access
    - Reduction in food insecurity rates in our communities
    - Increased access to fresh, healthy food options in former food deserts
    - Number of community members served by our Nodes
2. Environmental Impact
    - Tons of food waste diverted from landfills
    - Reduction in greenhouse gas emissions from our operations and supply chain

- Acres of land transitioned to regenerative agricultural practices

3. Community Health and Well-being

    - Improvements in diet-related health outcomes
    - Increased food literacy and cooking skills among community members
    - Level of community engagement in food system decision-making

4. Economic Empowerment

    - Number of jobs created, with a focus on living wage positions
    - Economic multiplier effect in local communities
    - Growth of local food businesses supported by our model

5. System Resilience

    - Ability to maintain operations during crises or disruptions
    - Diversity of food sources and supply chain redundancy
    - Community capacity for local food production

6. Innovation and Adaptation
   - New technologies or practices implemented
   - Successful pilot programs launched
   - Rate of knowledge sharing and best practice adoption across our network

By focusing on these metrics, we ensure that our growth is always aligned with our core mission and values. But collecting data is only the first step. The real power comes in how we use this information to drive continuous improvement.

Continuous Improvement: Always Growing, Always Learning

Our Nexus and Node model isn't set in stone - it's a living, breathing system that will evolve with our communities. We're committed to a cycle of constant learning and adaptation:

1. Regular Check-ins and Feedback Sessions: We maintain an open dialogue with every stakeholder, from farmers to families. Their insights drive our improvements.
2. Embracing Technology: We leverage data analytics and AI to identify trends and opportunities for optimization in our operations.

3. Academic Partnerships: We collaborate with universities and research institutions to study the long-term effects of our model on communities and ecosystems.
4. Annual "State of the Movement" Reports: These comprehensive reviews celebrate our victories, acknowledge our challenges, and chart our course forward.
5. Ongoing Training and Skill-sharing: We ensure everyone in our network is growing together, from culinary skills to business management.
6. Collaboration Across the Network: We facilitate knowledge sharing between different Nexus and Node locations, allowing innovations to spread rapidly.
7. Scenario Planning: We regularly engage in exercises to imagine potential future challenges and opportunities, ensuring we're always prepared to adapt.

This commitment to continuous improvement isn't just about making our model better - it's about creating a learning organization that can respond nimbly to the ever-changing landscape of food systems and community needs.

Balancing Growth with Our Core Values

As our movement gains momentum, we're not losing sight of what matters most. Our growth strategy is rooted in our values:

1. We're championing the small farms and local producers who are the heartbeat of our communities.
2. We're leveraging our collective buying power to push for fairer, more sustainable practices across the entire food system.
3. We're reaching out to existing food hubs, offering our model as a path to even greater community impact.
4. We're keeping our ears to the ground, ensuring that every decision we make is in service of the communities we support.

Becoming the Changemakers: Leading the Food System Revolution

We're not content with just running a successful network of food hubs. We're aiming to fundamentally reshape how America thinks about and interacts with its food system. Here's how we're positioning ourselves as the go-to experts in this new paradigm:

1. We're constantly learning, adapting, and refining our model based on real-world experiences.
2. We're sharing our knowledge far and wide, empowering other communities to join the revolution.
3. We're engaging with policymakers, advocating for the support and resources our movement needs to thrive.
4. We're bridging the gap between ivory tower research and boots-on-the-ground implementation, proving that transformative change is possible.

The Call to Action: Your Role in the Food Revolution

As we close this book, I want to emphasize that this isn't just my vision - it's an invitation for you to become part of something bigger than yourself. Whether you're a farmer, a policymaker, an entrepreneur, or simply someone who cares about the future of food, you have a crucial role to play in this revolution.

Here are some concrete steps you can take:

1. Start Local: Look at your own community. Where are the food deserts? Who are the local farmers struggling to find markets? What underutilized resources could be repurposed for food system improvement?
2. Build Connections: Reach out to others in your community who care about food issues. Start a discussion group, organize a community meal, or volunteer with local food organizations.
3. Advocate: Engage with your local government about food policy. Push for zoning changes that support urban agriculture, advocate for public investment in food hubs, or campaign for policies that support small farmers.
4. Educate Yourself and Others: Continue learning about food systems and share your knowledge. Host workshops, write articles, or use social media to spread awareness.
5. Invest: If you have the means, consider investing in or donating to initiatives that align with this vision. Your financial support can help pilot programs get off the ground.
6. Join or Start a Nexus and Node Initiative: Whether you're ready to lead the charge or want to support existing efforts, get involved in bringing this model to your community.

Remember, every great movement started with individuals who dared to imagine a different world. The innovators we've discussed throughout this book have already proven that transformative change is possible. Now it's your turn to add your voice, your hands, your heart to this movement.

Our communities are hungry - not just for food, but for connection, for purpose, for hope. The table is set. The work is waiting. And the time is now.

Welcome to the food revolution. Let's eat, let's grow, let's thrive - together. Take Action Now: [BOX]

1. Conduct a food system assessment of your local community, identifying strengths, weaknesses, and opportunities for improvement.
2. Organize a community meeting to discuss the potential for a Nexus and Node model in your area.
3. Start a "Food System Innovation" book club to continue learning and discussing these ideas with others.
4. Reach out to local farmers or food businesses to learn about their challenges and how you might support them.

5. Create a personal action plan for how you'll contribute to food system improvement in the next 6 months.
6. Share your commitment on social media and invite others to join you in this food system revolution.

The future of our food system - and our planet - depends on the actions we take today. Are you ready to be part of the solution?

Epilogue

STANDING ON THE SHOULDERS OF GIANTS

As we close the final pages of this book, we stand at the threshold of a new era in our food system. The Nexus and Node model we've explored together isn't just a dream or a distant possibility -- it's a roadmap for revolution, a blueprint for transformation that's already taking root in communities across the nation.

But this vision didn't emerge from a vacuum. Like any great movement, ours is built on the foundations laid by visionaries who dared to challenge the status quo, who saw possibilities where others saw only problems. As we prepare to carry this torch forward, it's crucial that we acknowledge the giants upon whose shoulders we stand.

Honoring Our Inspirations

Doug Rauch - Daily Table

Doug Rauch, the former president of Trader Joe's, showed us the power of innovative thinking in food retail with Daily Table. His hub and spoke logistics model and successful nonprofit grocery system proved that we can make good food accessible to all while building a sustainable business model. Doug's work reminds us that with creativity and determination, we can bridge the gap between surplus and need, turning potential waste into community nourishment.

Chef José Andrés - World Central Kitchen

In times of crisis, Chef José Andrés and World Central Kitchen have been beacons of hope, showing up in disaster zones with hot meals and unwavering commitment. José's work embodies the spirit of our movement -- the refusal to accept hunger as inevitable, the power of good food to heal and unite, and the incredible impact of passionate individuals rallying their communities. His example reminds us that in the face of seemingly insurmountable challenges, our resolve and our actions can literally save lives.

Robert Egger - DC Central Kitchen

Robert Egger's work with DC Central Kitchen laid the groundwork for our Nexus concept. By creating a robust, scalable model that addresses hunger, reduces waste, and provides job training, Robert showed us what's possible when we think holistically about food and community. His innovation proves that our food systems can be engines of empowerment, not just sustenance.

Project for Public Spaces

The team at Project for Public Spaces inspired us to view food as a cornerstone of vibrant communities. Their collaborative investment model, partnering with entities like Southwest Airlines, demonstrated how diverse stakeholders can come together to create lasting change. Their work reminds us that food isn't just about nutrition -- it's about creating spaces where communities can gather, grow, and thrive.

These are just a few of the countless innovators and community leaders who have lit the way forward. To all the food hub operators, urban farmers, community organizers, and tireless advocates we've met on this journey -- your dedication, your creativity, and your refusal to accept the world as it is has been our constant inspiration. You've shown us what's possible when we combine vision with action, when we refuse to let scarcity of resources become a scarcity of imagination.

The Call to Action: Join the Food System Revolution

As we honor these giants, we must also recognize the mantle they've passed to us. The Nexus and Node model isn't just an idea -- it's a call to action, a rallying cry for all who believe in the power of food to transform our communities and our world.

So, what's next? How do we take this vision and turn it into reality?

1. **Start Where You Are**: Whether you're a farmer, a chef, a student, or simply someone who cares about your community, you have a role to play. Look around your neighborhood. Where are the food deserts? Where are the opportunities for connection? That's where your journey begins.
2. **Build Connections**: Reach out to local food producers, community organizations, and potential allies. Remember, the strength of our model lies in its interconnectedness. Every conversation, every partnership, is a thread in the web of resilience we're weaving.

3. **Educate and Advocate**: Share what you've learned. Host community dinners, organize workshops, speak at local schools. The more people understand the power of local food systems, the stronger our movement becomes.
4. **Start Small, Think Big**: Don't be intimidated by the scale of the challenge. Even small actions -- starting a community garden, organizing a farmers market, volunteering at a food bank -- can have ripple effects beyond your imagination.
5. **Embrace Innovation**: Technology, urban agriculture, new distribution models -- be open to ideas that can amplify our impact. The future of food is being written right now, and you can be one of its authors.
6. **Demand Change**: Engage with local policymakers. Push for policies that support small farmers, enable food hubs, and prioritize community food security. Your voice matters.
7. **Stay Connected**: This book is just the beginning. Stay tuned for more resources, connect with others implementing the Nexus and Node model, share your successes and challenges. We're in this together.

Remember, every great movement started with individuals who dared to imagine a different world. The giants we've honored didn't wait for permission to change the world -- they simply began, with whatever resources they had, wherever they were.

Now it's our turn. The seeds of a revolution have been planted. With this book, you hold the blueprint for nurturing those seeds into a flourishing, resilient food system that can weather any storm.

The question isn't whether we can change the world. The innovators we've discussed have already proven that we can. The question is: will you join us? Will you add your voice, your hands, your heart to this movement?

Our communities are hungry -- not just for food, but for connection, for purpose, for hope. The table is set. The work is waiting. And the time is now.

Welcome to the food system revolution. Let's eat, let's grow, let's thrive -- together.

Made in United States
North Haven, CT
22 May 2025

69129473R10065